YOUR GENETIC ADVANTAGE

YOUR GENETIC ADVANTAGE

A Guide to Understanding Your DNA and Empower Lifelong Well-being

STEVE N. CONE

DEDICATION

To all those who have ever felt shame, guilt or inadequacy due
to society's unrealistic standards for weight and health,
Who only to see their efforts deemed failures through no fault
of their own.
This book is for you. Within these pages, I aim to remove the
self-blame you do not deserve and shine light on the invisible
genetic factors too often ignored.
It is my sincere hope this guide helps liberate you to
recognition of the unique genetic advantage within.

CONTENTS

PREFACE

Since first glimpsing the double helix structure of DNA over 60 years ago, scientists have made tremendous strides in decoding our genetic blueprint. Yet for many, the inner workings of our genes remain an elusive mystery. In these pages, I aim to demystify the field of genetics and illustrate how understanding your own DNA sequence can empower you to proactively support your health and well-being.

For most of my life, like so many others struggling with weight issues, I blamed myself for why diets repeatedly failed me no matter how strict I was with restricting calories or increasing exercise. No matter what approach I tried, the pounds never seemed to budge permanently and would often return with full force once I let up even a little on my efforts. It left me feeling like a failure and believing something must be inherently wrong with my willpower and discipline. Little did I know at the time how strongly influenced my genetic code was shaping that experience.

It was only after undergoing DNA testing in my late 30s that everything suddenly started making sense. The results revealed variants in several key obesity-risk genes that predispose individuals to store fat more readily and have a naturally slower metabolism. Learning this pivotal information was life-altering - it removed the immense self-blame I'd unfairly carried for so long and allowed me to understand my condition from a physiological perspective rather than a moral one. Perhaps most importantly, it filled me with hope that if I could strategize how to work in harmony with my genes rather than against them, long-term weight management may finally be within reach.

This insight inspired me to dive deep into the emerging field of genetic and nutritional medicine through my studies. I wanted to uncover practical solutions for people who, through no fault of their own, inherited genes making weight loss an ongoing challenge even with concerted lifestyle modifications. It became my mission to help empower others with personalized tools tailored to their unique DNA influences, avoiding the restrictive *"one-size-fits-all"* diets that so often backfire long-term.

Through years of dedicated research as both a medical practitioner and patient exploring this underrated topic, I aim to provide you all that I've discovered within these pages. My hope is that this book grants you the insider knowledge on which genes impact metabolism, cravings and other factors related to weight so you too can take advantage of your genetic profile rather than struggle endlessly against it and as well, feel empowered as an active participant in your healthcare decisions. Let this guide help you recognize your genetic tendencies not as a limitation but an opportunity for individualized optimization.

Most importantly, I share my takeaways on how to construct holistic, sustainable lifestyle habits in synergy with your biology for lifelong wellness, resilience and peace of mind moving forward.

Dr. Steve N. Cone

OBJECTIVES

Specifically, The Key Objectives Of This Book Are To:

1. **Validate Readers' Experiences** by showing genetics can legitimately increase obesity risk through no fault of their own. This provides an important perspective shift.

2. **Eliminate Misconceptions** that weight is solely determined by willpower and personal responsibility. Offering biological context can encourage more self-compassion.

3. **Empower Readers** by translating complex science into clear, actionable concepts they can apply to their own health journeys. Knowledge, as they say, is power.

4. **Inspire Hope** that even with genetic challenges, balanced lifestyle choices can still significantly impact health and weight outcomes over time. Non-scale victories matter as much as numbers.

5. **Spread Inclusiveness** by normalizing a variety of body types and health relationships. Beauty is diverse - it's about how one feels impacting the world.

6. **Build Community And Alleviate Isolation.** Many secretly battle weight issues. Sharing experiences forges kinship and mutual support.

INTRODUCTION

We've all struggled with our weight at some point. For many of us, it's an ongoing battle to maintain a healthy lifestyle and avoid gaining those extra pounds. But have you ever wondered if some people have it harder than others through no fault of their own? Recent research shows that genetics do play a role in our weight and our ability to lose or maintain it.

As someone who has to watch my diet more closely than others seem to, I was interested to learn that certain genes could be increasing my risk of obesity. Our genes determine a lot about our bodies - things like eye color, hair texture, even risk factors for certain diseases. So it makes sense that genetics would also influence how our bodies process and store fuel (food). Researchers have identified several genes associated with higher obesity risk, impaired satiety (feeling full), and lowered metabolic rates.

For me, I've noticed that even when carefully following my calorie goals, it's tough for the pounds to start coming off. On the flip side, I have friends who can seemingly eat whatever they want without consequence. At first I was a little jealous, wondering what their "secret" was. Now I realize their bodies may process food differently than mine due to genetic differences.

That's not to say their genetics are a free pass for unhealthy habits. We all still need to nourish our bodies with nutritious whole foods and make fitness a priority. But being aware of genetic predispositions can help provide some perspective during the inevitable stumbles. It's frustrating to feel out of control over your own weight, so understanding potential biological influences offers a sense of comfort. You're not a failure - your genes just give you an extra challenge compared to others.

So if you've ever felt like the weight loss battles are disproportionately stacked against you, take heart. Recent science shows you're not alone and it's likely not entirely your fault. Armed with knowledge about genetic obesity risk factors, we can have compassion for ourselves and others on a similar journey. Our environment and choices still matter greatly, but acknowledging biological influences may alleviate some of the self-blame often associated with weight struggles. You've got this!

1

You Are Not Alone

This chapter introduces the topic of genetics and obesity. We will discuss how many people struggle with their weight through no fault of their own, as genetics play an underlying role. Research showing certain genes can increase obesity risk are summarized in this chapter. A personal anecdote is shared about the author's own challenges losing weight despite diet and exercise.

For as long as I can remember, I've struggled with my weight. No matter how diligently I watched what I ate or exercised, the number on the scale always seemed stubbornly stuck. I was embarrassed by my body and constantly berating myself for lacking willpower or self-control.

Like most people, I blamed myself for every pound gained. "If only I didn't eat that slice of pizza" or "why didn't I go to the gym today?" But deep down, I knew others around me didn't seem to have the same issues. My friends could indulge in junk food and skip workouts without consequence. Meanwhile, I gave 110% effort and still lost nothing to show for it.

It wasn't until recent years that science started to validate my struggles as possibly out of my complete control. Groundbreaking research is revealing that genetics may play a larger role in weight than previously thought. Certain genes have been identified that can significantly increase an individual's susceptibility to obesity.

Knowing this new information has been life-changing for me. No longer do I solely blame every setback on personal failings. I now understand my body may be fighting an uphill battle from the moment I was conceived due to genetic influences outside my power. Doesn't that offer a sense of relief and reassurance for all the hours agonizing over each calorie?

You might be wondering how genes could cause weight gain, since we all ultimately consume the same number of calories in or out each day, right? Well, as with many aspects of our bodies, genes determine a lot about how our weight is regulated at a cellular level. They affect processes like metabolism, satiety, hormone function and more - all of which impact our ability to lose or maintain a healthy weight.

For example, certain genetic variants reduce production of a key satiety hormone called leptin. This means those of us with these variants never feel truly full, so we're more prone to overeating without realizing it. Genes also influence how efficiently our bodies burn fuel through metabolism. Someone genetically predisposed to a sluggish metabolism has to work much harder to burn the same amount of calories as their faster-metabolizing friends.

So in reality, not all calories are created equal depending on our unique genetic makeup. What's a simple choice for some *"I'll skip dessert to cut calories"* could require Herculean willpower for those with specific obesity- linked genes. My struggle was written in my DNA before I ever had a chance to fight against it.

Knowing these truths has allowed me to stop beating myself up over every minor slip-up. Instead of cruel self-criticism, I show myself compassion for the hand I was dealt biologically. I hope sharing my story provides solace for anyone else feeling they've struggled alone. You are not to blame. Let's walk together in empowerment rather than shame.

-*The Struggle is Real: My Journey with Weight and Genetics*

Like so many others, my battle with the scale began early in life. As young as 12 years old, I can remember feeling "bigger" than my peers and dissatisfied with my body. Fad diets and punishments seemed to do little but leave me feeling like a failure time and time again. By high school, I was at my heaviest – over 180 pounds on a 5'5" frame. The weight took a toll both physically and emotionally.

I tried every quick-fix under the sun after that – from meal replacements to extreme calorie cutting. Nothing sustainable seemed to work for losing more than a few pounds short-term. Meanwhile, friends effortlessly maintained their thinner figures without vigilant counting or deprivation. Why was this such a relentless struggle for me?

As an adult attempting to start a family in my late 20s, my weight had become the primary source of distress. Still averaging over 30 pounds more than my "ideal weight," I worried excess pounds could impact fertility or my ability to be active with future kids. So when a medical issue sent me to seek help from an endocrinologist, I grasped the opportunity to uncover what was hindering my progress.

With a panel of genetic tests and physical examination, my doctor's suspected I had an underlying predisposition all along- and she was right. Bloodwork revealed SNPs on several obesity-associated genes including FTO, MC4R, and PCSK1. Further tests pointed to sluggish metabolism and blunted satiety signals. Armed with this context, my struggles began making much more sense.

My biology was primed for excess fat storage and hunger cues on a level hardwired since conception. What had seemed like failure was now cast in a new, compassionate light as misfortune beyond my power. Suddenly, I felt a weight lift beyond numbers—relief from years spent blaming and berating myself harshly. If others could lose by cutting calories alone, maybe something else was at play restricting me.

This revelation set me on a path of learning how to peacefully partner with my prewired tendencies through small sustainable adjustments. With a doctor's guidance tailoring diet, movement and mindset—slowly but surely, my relationship with food and body began transforming in a kinder direction. I now understand the secret of wellness lies not in victories over our fallibilities but acceptance of our humanity in all its variegated forms.

My hope in sharing is empowering others stuck in the same cycles of self-loathing to see themselves—and release unrealistic notions of what wellness demands from diverse people and lives. There is lightness available when we lift burdens of judgment from one another and ourselves. Our varied genetic blueprints deserve compassion, just as we all do in this shared existence.

2

Our Genetic Blueprint

In this chapter will dive deeper into the science behind genetic influences on weight. It will explain how genes determine many bodily functions and traits. Key genes linked to obesity through processes like metabolism and satiety are profiled.

Before delving into specific genes, it's important to understand the basics of how genetics determine our bodies at the most fundamental level. You may be familiar with DNA - that *double-helix* molecule found in nearly every cell which contains our genetic instructions.

DNA is made up of nucleotide bases assembled in a specific sequence that spells out our genetic code. This code gets expressed via messenger RNA and proteins to form all our physical traits. For example, genes dictate everything from eye color to disease risk through precise biological pathways. Genetics serve as the blueprint that build us from formation in the womb outwards.

When it comes to weight regulation, our genetic code contains instructions for numerous metabolic processes. Genes help control how our cells convert food into energy through metabolism. They influence appetite regulation via satiety hormones like leptin and ghrelin. Genetics even impact our fat storage through specialized fat cell genes. All of these systems working in concert determine our baseline body composition and weight trends.

Now Let's Look Into A Few Specific Obesity-linked Genes Researchers Have Identified:

FTO Gene: Carrying variants of the FTO gene is the strongest single genetic factor for obesity risk. Studies found those with variants are 70% more likely to be overweight. FTO influences metabolism and energy balance in the brain's appetite control center.

MC4R GENE: MC4R gene variants are the second most prevalent obesity gene. They reduce function of the MC4R protein, a receptor that transmits satiety signals from leptin to the brain. This predisposes to overeating.

POMC & PCSK1 Genes: Variants impacting these genes decrease production of appetite-regulating peptides like Alpha-MSH. This raises hunger and lowers metabolic rate, increasing fat storage susceptibility.

While single genes have modest effects, the more obesity variants each person carries compound their genetic influence. Additionally, genes interact with lifestyle and environment in complex ways still being uncovered. But isn't it fascinating how much science is now revealing about bodies we once felt so out of control over? With knowledge comes empowerment.

−Genetics and the Biology of Weight Regulation

Now that we've established how genetics can influence weight outcomes, let's dive into the complex biological mechanisms at play. Every pound gained or lost stems from cellular processes finely tuned by our DNA. Metabolism, hunger cues, fat cell activity - genes help orchestrate it all through interconnecting pathways.

At the foundation lies our metabolic rate, a measure of how many calories our bodies burn daily at rest. Components like thyroid function, muscle mass and mitochondrial activity impact this basal number enormously. Genetic variants can slow metabolism by up to 200 calories per day, equivalent to a full meal's worth!

Satiety also depends on genetic expression. Key hormones provide signals between digestive tract, pancreas, adipose tissue and brain to regulate appetite. Leptin produced by fat cells tells our brain "I'm full, stop eating." But for some, genes yield less leptin receptors impairing this signal.

Other hunger hormones impacted include ghrelin, the "empty stomach" trigger made before meals. People with certain ghrelin-encoding genetic variants tend to ignore these fullness cues, inadvertently overconsuming without noticing.

On a cellular level, adipose or fat tissue comprises specialized genes allowing triglyceride storage in adipocytes. The more adipocytes one has, the more excess calories can get locked away rather than used for fuel. Genetic makeups conducive to rapid fat cell division heighten predisposition to obesity.

The intricate dance of these various bodily systems illustrates how multi-factorial weight regulation truly is. While environment plays a role, genetics predetermine our default biological tendencies from the get-go. No level of restriction from willpower alone could overcome certain preprogrammed biases.

Armed with this science, we gain insight that liberates from self-blame. Our genetic code is simply the starting foundation from which lifestyle shifts can mitigate embedded risk factors through nurture too. With knowledge, acceptance and empowering habits, well-being flourishes.

3

NURTURE vs NATURE

While genetics are acknowledged as an obesity risk factor, this chapter emphasizes that lifestyle still greatly matters. A balanced perspective is provided, noting that genetic predispositions do not equal an automatic pass for unhealthy habits. The interaction between our genetic blueprint and environmental stimuli will be explored here. Tips offered on mitigating genetic risk through optimizing diet, stress, sleep and exercise.

Now that we've explored the role of genetics in weight regulation, let's clarify that while biology matters - our behaviors and choices still hold immense sway as well. After all, genetics are simply our inherited tendencies, not outright determinants of destiny.

Environmental stimuli like the foods we eat, how active we are, stress levels and more interact dynamically with genetics each moment. Take the example of someone genetically prone to overeating due to low satiety signals. If that person lives in an obesogenic environment surrounded by calorie-dense fast foods and screens, their genetic predisposition will likely cause greater weight gain than someone practicing moderation despite the same genes.

Life is all about balance. Instead of seeing genetics as an excuse to throw in the towel, view your personal risk profile as extra motivation to double down on healthy habits. Prioritize home-cooked whole foods, stay hydrated, minimize stress, and aim for 30 minutes activity daily - all of which aid weight maintenance regardless of our DNA hand.

Social support also provides a genetic buffering effect. Leaning on compassionate loved ones during hard times helps alleviate stress linked to metabolic dysfunction. And don't go it alone by only focusing on weight - take care of your whole well-being through hobbies, community and purpose as well.

Finally, love your body as is and don't let a number define your worth. Focusing so much mental energy on weight often backfires by inducing unhealthy restriction or binge cycles. Have your health goals as a side benefit, not the main purpose, of living joyfully each day.

More Tips On Mitigating Genetic Risk Factors Through Lifestyle Habits:

Nurturing Wellness – Optimizing Your Health Through Empowered Living

By now, we've explored the influence genetics can have on weight regulation. However, knowledge is only power when applied constructively. In this chapter, I want to share more specific lifestyle strategies you can use to work in partnership with your DNA, rather than feeling ruled by forces outside your control.

How do we nurture well-being when working within biological limits outside our control? This is where lifestyle empowerment comes in through small consistent habits.

-Diet plays a huge role it is crucial for balancing appetite signals. Focus on satiating whole foods like oats, lentils, nuts/seeds, salmon, Greek yogurt and avocado that steady blood sugar. Aim for 30+ grams fiber daily from veggies or fruit to promote digestive health and fullness. Eating regular mini-meals every 3-4 hours also prevents plunges in glucose. Focus on lean proteins, fiber-rich complex carbs, healthy fats and nutrient-dense produce at each meal.

These satiating whole foods stabilize blood sugar and optimize gut health to help maximize satiety. Aim for 5 servings of fruits and veggies daily minimum.

-Hydration lightens the metabolic load. Sip water throughout the day and avoid calorie-packed beverages which add up fast without much nutrition. Tea and coffee in moderation are fine too with some benefits. Staying hydrated also impacts metabolism and food processing. Drink half your body weight in ounces of water daily. When thirsty signals are confused for hunger cues, overeating is more likely. Water is also nature's appetite suppressant.

-Be 'Activity-wise', aim for a mix of cardio, strength and mobility daily even if just 10-20 minutes broken into short sessions. Physical activity not only burns extra energy, it expands lung capacity and boosts mood-regulating endorphins. Establish a regular movement routine that works for your schedule. Even short 10 minute bursts of activity add up. Simply parking further, taking walking meetings or playing upbeat music while cleaning all count. Gentle exercise boosts good hormones to combat genetic risk.

-Then there's stress management through whatever practices helps **you** unwind - yoga, nature walks, painting, baking etc. High cortisol opposes fat loss efforts by spiking appetite and insulin, placing extra demands on the body and challenges weight loss. Calm is key.
Minimize stress through hobbies, socializing, meditation, journaling - whatever gives you an outlet. Laugh with loved ones daily for a metabolic spike too.

-Sleep schedules are equally restorative for both physical and mental strength. Quality slumber optimizes growth hormones, repairs tissues and recharges cognitive function for the following day's challenges. Sleep quality holds metabolic significance as well. Aim for 7-9 hours as genetics disrupt when tired. Develop a bedtime ritual without screens an hour before sleeping to unwind fully.

Implementing this holistic lifestyle framework steadily over time and being mindful of these empirically backed lifestyle habits helps counterbalance genes outside our control. While no one strategy provides a "cure", consistency gives your body the best chance at health regardless nature's blueprint under the hood. Progress occurs one small conscious choice at a time. You've got this!.

THE BOTTOM LINE

The bottom line is we all deserve kindness - toward ourselves and others. Genetics offer context, not condemnation nor excuse. With self-care and perseverance, you have power over your path regardless what hand you were genetically dealt.

4

WHAT IT MEANS FOR YOU

Understanding Your Personal Risk Profile

This chapter should help readers determine their own genetic risk profile. A questionnaire is provided to gauge potential inherited influences. Advice is given on discussing genetic test results with a doctor.

Now that we've covered the scientific foundations and healthy living strategies, it's time to help you better understand your own genetic makeup. While a full genome sequencing remains quite expensive for most consumers, direct-to-consumer testing panels for common obesity risk variants are affordable options available online.

These genotyping tests analyze your DNA for specific genes we've discussed, such as FTO, MC4R and others. You provide a saliva sample that is shipped to a lab for processing. Within a few weeks, you'll receive your unique genetic report highlighting any obesity risk variants detected.

It's important to note, as with all genetic testing, results require proper context. Having a single or even multiple variants does not mean weight gain is definite or predetermined. Think of genetics more as providing an obstacle that requires more effort to overcome, not a locked destiny. Your actions still matter greatly as we discussed in Chapter 3.

I recommend reviewing testing reports with your regular doctor. They can best explain findings and ensure you appropriately consider both nature **and** nurture. Don't become overly defined or limiting based on a test. Lifestyle makes a difference!

Remember, your value lies not in any number on a scale or lab report. Results offer perspective, not judgment. See challenges simply as hurdles versus characterization of self-worth. Knowledge fosters empowerment when handled constructively.

With a self-care routine and community, you have capacities to achieve health dreams independent what DNA may say about tendencies. Release perceived failure from process by appreciating non-scale victories alongside small successes over time. You've got this - let's maintain a growth mindset moving forward!

Unlocking Your Potential: The Insights of Personal Genetic Testing

We've quite understood the interplay between genetics and environment, you may be curious about your own biological predispositions. While our individual genotypes cannot be changed, gaining clarity provides empowering context for self-care.

Direct-to-consumer genetic tests analyze common obesity-linked gene variants through an at-home cheek swab sample. Within a few weeks, board-certified genetic counselors help interpret your results. While single SNPs have modest effects, cumulative risk elevates with additional predisposing variations.

Results aren't pass-fail grades but spectra - the more detrimental versions someone inherits, the higher their probability of weight regulation challenges. However, as we've explored, potential is not a fixed destiny with lifestyle optimization.

Its vital tests are never viewed as passports toward fatalism. Rather, understanding your personalized tendencies equips compassionate self-awareness. Whole numbers alone don't define worth, knowledge supports personalized goal-setting and makes slip-ups feel less catastrophic.

Some find test results validating after a life blamed unfairly for conditions outside control. Others realize predispositions warrant even more self-care. Either way, de-stigmatized discussion generates empowerment over self-sabotage.

Regardless of outcomes, maintain perspective. Genetic contributions comprise just one thread in the tapestry of our rich humanity. Physical diversity deserves the same respect as any other. So as always, uphold kindness - for yourself most importantly. Your inherent dignity transcends all metrics.

If interested in exploration, I suggest consulting professionals throughout. By honoring both science and psyche, genetic literacy elevates well-being for people of all constitutions. Light and wisdom await within.

5

Taking Control

In this concluding chapter, you will be empowered with a positive mindset for managing your weight goals. While genetics cannot be changed, control can be had through lifestyle modifications. Success stories inspire how others have overcome genetic challenges through committed wellness routines. This chapter reminds you that 'You are not alone' in the journey and have all the tools within to meet your health goals.

Thriving Within and Beyond Numbers

By now you've gained an understanding of genetics' influence as well as practical steps to support your well-being. In this culminating chapter, I want to help refocus your mindset on the holistic joys of living well beyond any number or test outcome.

A long-term sustainable lifestyle focuses less on temporary goals and more on daily small acts of self-care. Delight in nourishing body and soul through movement you love, bonds with loved ones, creativity, nature and more. Weight shifts occur gradually alongside life's natural ebbs and flows this way.

Learn to appreciate non-scale goals just as much as any number – like boosted energy, healthier sleep, stronger confidence, increased patience or stronger community ties. Note personal progress through gratitude journaling too rather than harsh analysis alone.

Envision your best self and seek like minded groups who celebrate diverse bodies and accomplishments. Develop this supporting tribe for consistent encouragement through individual ups and downs alike. We all need compassionate allies as genetics intersect lifestyle in unpredictable ways.

Finally, pay it forward with what you've learned. Loving yourself includes loving others by spreading awareness of genetic nuances versus societal assumptions. Together, empowered change happens person by person through open dialogue over sensationalism and stigma alone.

Your journey holds meaning regardless of terrain traversed or destination. Stay rooted in purpose, perseverance and community. With empathy, empowerment replaces victimhood. You've inherited strengths as well as tendencies - now fly boldly within the skin you're in! I believe in your greatness and ability to thrive wherever life's path may lead.

Nourishing Connection: The Importance of Community
This journey of understanding our genetic tendencies and nurturing wellness is deeply personal yet not meant to be traveled alone. In this chapter, I want to stress the vital role that community plays in sustaining life changes and cultivating self-worth disconnected from scales or diagnoses.

Surrounding ourselves with like minded souls makes challenges feel less isolating and more surmountable through shared wisdom and accountability. Online forums or local in-person groups provide invaluable camaraderie for those privately experiencing self-perceived differences quietly for years.

Within the community there exists a reciprocal understanding that no person struggles alone. We all navigate obstacles society deems imperfections. So rather than judging, lift one another with empathy, validation and compassion as equals deserving joy, just as we wish for ourselves.

Find purpose beyond health by cultivating gifts giving back brings. Volunteer in causes supporting others' wellness like tutoring children, meal prepping for sick neighbors or raising awareness of genetic diversity normalize through art. Pour energy outward and watch fulfillment overflow in return.

Pursue hobbies purely for their own rewards too like dancing, crafting, learning an instrument or game nights with friends. Relationships sustain us far beyond any fleeting physical changes or numbers triggering self-criticism. Your worth was never tied to such trivial things.

Each mind, body and spirit is a novel, irreplaceable work of wonder deserving appreciation for all seasons contained within. Release control over what you cannot change and instead embrace the privilege of each sunrise. May your days be filled by nurturing loved ones, spreading encouragement for all travelers on parallel paths, and overflowing with contended presence amidst life's ebb and flow. You have so much to savor!

CONCLUSION

And so my friends, we've come to the end of our journey together exploring the intersection between our genetic predispositions and lifestyle empowerment. I hope you've gained invaluable perspective into the complexity of weight regulation as well as tools to champion your long-term health and well-being. The overarching aim of this book is to educate and empower readers about the role of in weight regulation.

Your DNA is the code of life encoded in every cell. By understanding how it shapes your health destiny, you gain self-awareness and empowerment over both risks and gifts inherent in your genetic blueprint. This book has shown how genomic literacy is crucial for each lifestyle choice and medical decision moving forward. With curiosity and compassion, embrace your genetic individuality while also recognizing our shared human nature. Take control and optimize your genetic advantage - your unique path to thriving wellness of body and mind.

By providing scientifically-backed information in an approachable way, I can be assured of alleviating feelings of shame, self-blame and isolation that often accompany weight struggles.

Most of all, I hope this work has provided compassion - for yourself and others walking parallel roads. We are so much more than any characteristic or number could define. Each ensouled vessel deserves grace, especially when dealing with pre programs beyond willful control. May you lift each other gently as you sustain self-care with community's loving support.

Go now in peace and purpose. May your days be filled with presence, connection, laughter and light. I am forever grateful you joined me on this mission of empowerment through wisdom. Our shared humanity transcends all metrics. You've got this - now fly joyously within the skin you're blessed to inhabit!